I0476005

Integrated Project Planning and Construction Based on Results

Williams Chirinos

Published by Williams Chirinos, 2020.

INTEGRATED PROJECT PLANNING AND CONSTRUCTION BASED ON RESULTS

First edition. October 15, 2020.

ISBN: 978-1386470090

Written by Williams Chirinos.

Table of Contents

Introduction

The project life cycle is the series of phases defined by an organization to develop a project from the idea to its operation. There are two main types of project life cycles; predictive or adaptive, and any combination between these two is called a hybrid project life cycle.

Predictive life cycles are serial or sequential construction projects that create tangible products such as schools, hospitals, bridges, buildings, refineries, roads, and others. They are also called plan-driven projects because the project team has a plan before going into the field of construction.

Adaptive life cycles are iterative or incremental projects that create intangible products such as software applications, methods, procedures, templates, policies, standards, intellectual properties, and others. They are also called change-driven projects, and agile is increasingly implemented to manage this type of projects properly.

This book focuses on predictive life cycle projects, projects that have a field construction stage of building tangible assets. Specifically, the content refers to the execution phase of projects, when you need to prepare a baseline plan during the planning phase to go to the field of construction to build the asset and finally deliver it to the customer. The construction phase is the target because it is here where large quantities of the project budget are spent. Regardless of the industry, most of the time, construction projects have difficulties being successful, even some of them go into claims and dispute resolution situations.

This book serves as a guide for all those seeking to improve project delivery using new approaches and methodologies to plan and control the construction phase of projects more effectively and efficiently.

Chapter 1 presents the conditions for successful projects with a real case study project. This chapter also discusses the interaction of critical stakeholders and factors to achieve high-level project team performance, such as support, collaboration, trust, and communications.

Managing projects means managing people, and chapter 2 contains seven power skills – commonly known as soft skills (decision making, conflict resolution, problem-solving, negotiation, leadership, alignment, and optimistic bias) for project practitioners. Power skills that specifically, project leaders need to grow and develop during their professional careers.

Chapter 3 describes the basic and advanced project management levels and introduces the result-based concept and technique. The potential implementation of the result-based concept in complex projects is also discussed in this chapter.

The description of scope baseline, work breakdown structure, resource allocation, cost and schedules estimates, and baseline plan are presented in chapter 4 using an example of a House Construction project to explain the project planning phase based on results - deliverables and work packages, instead of activities or task. The approach of focusing on deliverables and work packages, instead of focusing on activities or tasks requires a change in the way of thinking about projects.

Chapter 5 starts with the importance of procurement as a requirement before construction. Then, the chapter describes the construction phase using the result-based concept, which means focusing on deliverables and work packages (instead of activities or tasks). This chapter also explains the concept of earned value with a case study project and the use of earned schedule, and it ends with the typical content of project status reports during construction.

INTEGRATED PROJECT PLANNING AND CONSTRUCTION
BASED ON RESULTS

Change management is addressed in chapter 6, with the comparison between the baseline plan and the actual data that comes from the field of construction using variance analyses. Changes likely happen during construction; therefore, this chapter also addresses the re-baseline due to significant variances and the risk management processes.

Chapter 7 explains the project completion stage, commissioning, and the lessons learned as the final steps of construction projects to ensure the product or service is ready to be issued to the customer in full compliance with the agreed specifications and requirements.

As a result of the previous chapters, chapter 8 presents a summary of the integrated planning and construction methodology and a cloud-based software application called PEMS (Planning and Execution Methodology Software) to start the implementation of the ideas presented in this book and start making a difference in your construction projects.

Successful Projects and Stakeholders

Successful Projects

A project is a unique and temporary process undertaken to produce agreed results. The agreed results could be products or services because projects not only deliver products such as transportation road systems or buildings, but also services such as technical reports or pipeline system hydro-tests.

It is generally accepted in the project management community that a project is considered successful if, at the completion stage of development, the project ends: delivered on time, with a final actual cost on or below budget, and in full compliance with the technical and regulatory requirements. Under the previous definition, statistics indicate that around two out of three projects have difficulties meeting all the previous three successful criteria. This situation is the reason why there is still a need for new approaches and methodologies to plan and control construction projects more effectively and efficiently in the project and construction management communities, regardless of the project size, complexity, type, or industry.

Case Study of Project Success or Failure

Let us elaborate on a case study to determine if the following project was or was not successful.

A pipeline 12-inch diameter and 10-kilometer length that transports processed water from plant A to plant B arrives at its final life cycle after 30 years in service and needs to be replaced. The replacement cost is 20 million dollars and takes 2 years to complete. The project team realizes that plant B is going to be in operation for only 12 years more; after that, plant B is going to be decommissioned. Under this new condition, it makes sense to install a high-density liner inside the existing pipeline to extend its service time for 12 more years, instead of performing a replacement. With this new scope of work, the project's new cost estimate is 6 million dollars, and the new duration estimate is 8 months. It was approved to install the liner instead of replacing the pipeline, and the installation of the liner starts. During the construction phase, a couple of pinholes occurred during the hydro-static tests to check the mechanical strength of the pipeline, before installing the liner. It was challenging to identify the location of the pipeline failures, and the construction team spent more time and resources to replace the two failure sections. In the end, the final total cost of the project was 8 million dollars and took 1 year to finish it, and the pipeline returned to operations according to the requirements. The question is, was the project a success or a failure?

Some people believe the project was a failure because although the requirements were met, the project overspent 2 million dollars and experienced a delayed of 4 months. Other people believe the project was a success because instead of spending 20 million dollars with a 2-year duration, the final cost was 8 million dollars and took 1 year, with a more technical suitable solution. This situation shows you that sometimes project success or failure depends on the point of view and the criterion of the observers.

Project Team Performance

Performance is the process of fulfilling the intended purpose. Project team performance happens when the people working on the project, fulfill the purpose of their contribution to the project objectives (scope, schedule, cost, and quality), regardless of their knowledge, skills, or discipline.

As project managers, it is crucial to encourage the execution of duties with high performance to each one of our project team members. This approach is a difficult task because leadership and motivation are involved, and one of the most important elements here is creating an appropriate working environment for engagement. Different factors contribute to a high-performance environment, such as support, collaboration, and trust. Situational examples of each factor are as follows:

Support: being supportive of our project team members is one of our primary functions as project leaders. I was working on a bilingual storage facility project, the design engineering team was in one country, and a second engineering team was in another country, the latter country having the storage facility construction scope. The foreign engineering team was supporting our home design team, making sure the project met the technical standards of both countries. Due to the two different languages, an opened line of communication and translation of critical documents was set up in both countries to ensure consistency and effectiveness in the communication.

Collaboration: this is a virtue attitude cultivated by the reciprocal give and take approach. I was assigned to a process plant design project that was 50% complete. During the first project review meeting, I realized that there was much discussion among the team members regarding design issues; moreover, most of them were waiting for technical information to advance the design. The situation improved when we create a follow-up procedure using an issue log template to keep track of the missing technical information. Besides, we stopped sending emails not related to the design issues and focused on sending emails only to communicate a technical decision or to deliver specific design information.

Trust: it is the golden factor in a high-performance team environment; it is also hard to build and easy to lose. I was involved in a project with a significant fabrication component. The fabrication company gained the trust of my project team because they were proactive in attending some inconsistencies on the mechanical drawing package, and they did what they said they would do. Trust is something that you gain when you walk the talk, and when you live up to your values.

Stakeholder Management

A stakeholder is a person or entity related to the project during the project life cycle. Stakeholders influence the outcome of each project, and as project managers, you need to identify your stakeholders to communicate with them. You must request or provide the appropriate amount of information to the stakeholders at the right time to ensure successful project completion. Examples of stakeholders are; project sponsors, project team members, top management, vendors, consultants, contractors, final users, the public, professional associations, the government, financial institutions, and others. Project practitioners have realized the importance of the stakeholder management, and the Guide to the Project Management Body of Knowledge (PMBOK® Guide) has a chapter dedicated to Project Stakeholder Management to support project people with this not simple project component.

Stakeholder Communication

Let us classify the stakeholders as per figure 1 below. Communication is mainly bidirectional with Performers (project team members, subject matter experts, and the Project Management Office or PMO) and Suppliers (vendors, consultants, and contractors) because they advance the project through its life cycle until completion. On the other hand, communication is mainly unidirectional with Clients (sponsors, top management, and final users) and Influences (the public, government, and professional associations) to provide project information and get approvals to proceed with the construction phase of the project.

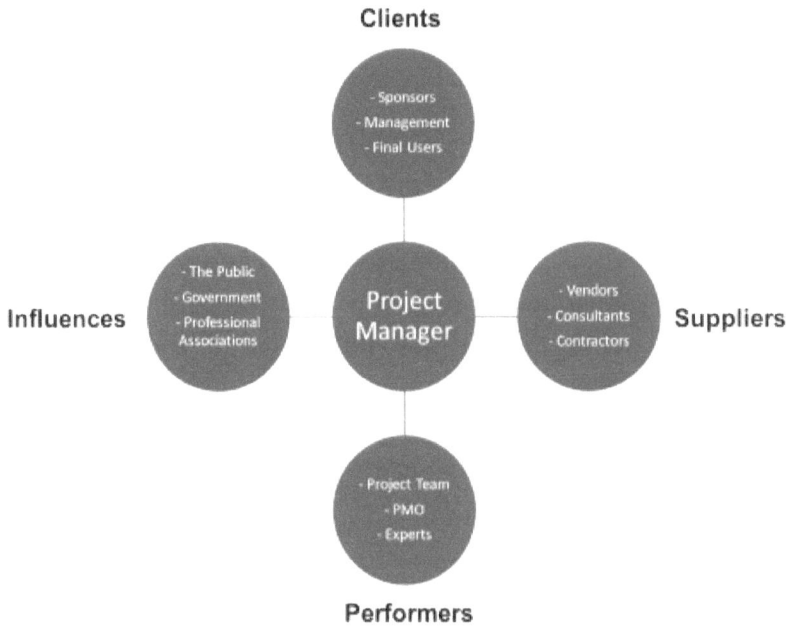

Figure 1. Stakeholder Communications in Projects

Importance of Effective Communications in Projects

Most of the project people would agree that one of the most paramount reasons why projects fail is because of poor communications among the stakeholders. Effective communication means the sender encodes and conveys the messages, and after a while, the receiver receives and decodes the messages. Critical aspects to consider during this process are; the encode and decode have to be on the same platform such as type and level of language used, the medium to convey and received needs to be adequate like a periodical project status report. Usually, the messages need to be confirmed by the sender or receiver asking questions to each other. The final objective in projects is to have a consistent communication system to answer the two most important questions among the stakeholders; 1) Where are we? (project percent complete) and 2) Where are we heading? (forecast of project final cost and completion date).

The management of stakeholders involves identifying whom they are, gathering their expectations, giving them the project information, and developing and executing strategies to gain their understanding and support. It is a delicate negotiation process, and as such, it is crucial to identify the common interest of each party. It is not about positions; it is about agreeing on the benefits that the project is going to bring to each party under the win-win approach. Remember, the project success depends on the goodwill that this is going to bring to the stakeholders, and the goodwill is conveyed with trust through an effective project communication system in place.

Project Sponsors as Stakeholders

A stakeholder is a person or entity related to the project. Project Sponsors are critical stakeholders because they usually represent the client or customer, and they have direct contact with the project manager, top management, and in some cases, with project team members. It is becoming a good practice that the sponsoring office is located nearby the physical location of the project team, and in Agile projects, the sponsors work as one more project team member. Sponsors are the ones that elaborate and approve the project charter; therefore, they know the required project scope of work, business case, expected benefits, general requirements, high-level risks, high-level expectations, and the success criteria. The latter shall be identified during the project charter creation, and it is usually related to time, budget, or quality. It is imperative to take the time to establish a trustful relationship with the project sponsor from the beginning of the project.

Reliable Data, Support, and Engagement

INTEGRATED PROJECT PLANNING AND CONSTRUCTION BASED ON RESULTS

A relationship based on trust with the project sponsor is established by exchanging reliable data, providing mutual support, and maintaining his/her engagement in the project.

Reliable data is information that comes from the field of construction as a result of the project execution. The data needs to be consistent and based on facts (instead of opinions). Useful data is the one that allows project managers to determine what is happening in the construction field, analyze the project performance, and implement corrective and preventive actions as needed. The outputs of reliable data are conveyed to the project sponsor regularly through steadfast project status reports.

When mutual support between the project manager and project sponsor exists, it is a good signal of trust. However, trust needs to be created for most of the first-time projects. To create trust, it is essential to living your values, to "walk the talk." One of the best references for a standard set of values to pursue is written in the Project Management Institute's Code of Ethics and Professional Conduct, follow these values, and you are going to be ahead of the game in creating trust with your project sponsors.

As project managers, we need to maintain the engagement of sponsors in the project environment because we need to make decisions to advance the construction of the project consistently. One way of doing this is to manage the relationship with project sponsors closely, this means to involve them in discussions, consult them when appropriate, and engage them when making tough decisions.

Teamwork and Collaboration

Teamwork is the combined effort of each member to achieve a collective result; here, the quote by Aristotle, "the whole is greater than the sum of its parts" applies perfectly. On the other hand, collaboration is working in alignment to create or produce something; here, the interactions go beyond the teamwork environment.

Difference Between Teamwork and Collaboration

Teamwork is critical in projects; it means that each one of the project team members must do his/her part of assigned work to deliver what the project is supposed to deliver as a whole. Once the project starts, members of the organization are selected to form the project performer team. Since expectations are high at the beginning, and the project team is focused on the scope of work definition, this is a valuable time to clear expectations and start building trust performing consistent actions towards the common goal. As time goes by, the project team members are going to have the opportunity to perform teamwork. They are not only going to do things right, but also are going to do the right things to produce the part of the project results under their responsibility, and in compliance with the specifications of quality and time.

Collaboration goes beyond teamwork; it is the result of interactions between the project team members and outside entities such as:

- Project sponsor
- Subject matter experts
- Material and equipment suppliers
- Sub-contractors
- Vendors
- Service providers
- Government regulators
- The public

Collaboration involves all the people affected by the project. Here, the most important thing is to get alignment among all stakeholders concerning the final project objectives, the reason for doing this is because not always do the outsider stakeholders have the same interest regarding the project objectives. For example, building a pipeline in the backyard of a private landowner; needs an agreement between the landowner and the project sponsor. Effective collaboration among stakeholders is obtained by building trust, and trust is built by establishing effective communications and commitments from start to finish.

Project Influences as Stakeholders

Project influences are stakeholders external to the organization and the project team. Influences are people or entities interested in projects positively or negatively. Usually, their opinions need to be taken into consideration because they address technical, business, or public related aspects. Examples of influences are the public, government, financial institutions, and professional associations. Let us explore some examples of these project influences.

The Public, Government, Financial Institutions, and Professional Associations

The public usually has high power and low interest in projects, but they are like a sleeping giant; if for some reason, their interests increase, they certainly have a strong influence on the project objectives. As an example, one hydraulic power generation plant project was progressing in a Central American country, when construction started the project team realized there was opposition to the project posted by the surrounding community villas. People from the villas frequently blocked the access roads to construction, and the company had to put the project on hold and initiated further consultation with the communities.

The government, represented by the regulatory bodies, permits the construction project to go ahead. The government has high power, impact, and interest in projects. As a result, it has significant influence as well. In Canada, a water pipeline project was in the design stage to replace the existing pipeline that transports water from a lake to a process plant when the government instructed the company to be off the lake in 10 more years. As a result, the company had to change the design to install a high-density liner inside the existing pipeline to extend its service for 10 more years, instead of performing a replacement.

Financial institutions have a high impact and interest in the projects they sponsor. The professional engineering association in Lima Peru was planning the replacement of it headquarter building and wanted to execute the 3-year project in one phase (a change from the original plan of executing the project in two phases). The bank told the engineering association that to receive the financial support, they had to continue with the original plan and execute the project in two phases because of the risk of finding a geological treasure during the excavations and having to postpone or cancel the project.

Professional associations usually have low power and interest in projects; therefore, they have a little influence on them also. The way professional associations may take part in projects is when they investigate any breach of the code of ethics. They act on a case by case basis, after someone reports a breach of professional misconduct or conflict of interest regarding the project.

Seven Power Skills for Projects

Decision Making

Decision Making is the process of making up our mind regarding several options. In project management, it is the process of selecting the most suitable solution to progress the project. During this process, a series of steps are executed to pursue an appropriate outcome, such as gathering information, elaborating the options, comparing the options, and choosing the one to follow.

Factors That Affect the Decision Making

Currently, People are bombarded with plenty of information. This condition affects an effective decision-making process because of the distraction factor. As an example, let's look at the email. There is no doubt the email is a powerful project communication tool, but sometimes people receive so many emails per day that they may feel overwhelmed. When this happens, people usually feel unable to think and adequately make critical decisions that are needed to progress the project. One critical recommendation to the project team is to have different folders with different priorities such as Projects, Engineering, Procurement, Fabrication, Construction, and others. When opening the inbox, people can read the subject (and maybe the first sentence) and move emails to the appropriate folder (even to the Trash Bin) so people can focus on the most important and critical problems that require deciding for immediate action. A practical recommendation is sending emails if a decision has been made, an agreement reached, or to record information.

Complexity also affects an effective decision-making process. As project leaders, pursue simplicity instead of complexity; keep things simple to keep people focused on what is essential. To illustrate this, when you have a project, you list all the activities and include them in a colorful schedule with hundreds or thousands of activities. Then when the project starts, people rapidly start executing activities, and usually one month down the road, the project is behind schedule. Instead, as leaders, focus on lower and medium level results that are going to get you where you want to be and act, don't be afraid of taking calculated risks. It is about making decisions, and applying a balanced-management criterion, avoiding micro or macro-management.

Timing is another factor that affects effective decision making. Time is the independent variable by excellence, and each decision is usually associated with time to respond. It is crucial to sense the time when a decision must be made; otherwise, there is a risk of diminishing the outcome of the decision. As an example, if a design error is identified during the construction stage, it means that the damage (e.g., more cost to purchase equipment under specs) is going to hit the project because of the appropriate decision to reject a drawing during the design stage was not taken.

Conflict Resolution

Conflict resolutions in projects are the steps taken to address opposed opinions or controversies among the stakeholders. The main idea is to promote climates in favor to reach agreements, where strong disagreements exist. It is vital to add value in the process of conflict resolution to minimize adverse effects on the project progress.

How to Arrive at A Valuable Conflict Resolution?

Project managers are the ones to mediate in conflict resolutions because they lead the project team. They have access to top management people, and they keep communication with service and material providers; this means project managers handle a high volume of information, vertically and horizontally. During the process of conflict resolution, it is essential to apply basic principles to look for fairness and positive outcomes.

The following steps are presented to help project people reach favorable conflict resolutions:

1. Address the problem as soon as it appears.
2. Have meetings with each member in conflict first, and then have a combined meeting or vice-versa.
3. Find out what had happened from different sources.
4. Look for and evaluate the facts.
5. Stick to the basic principles, because they do not change, the situation is what changes.
6. Ask specific questions to find out the interests of each people involved in the disagreement.
7. Find out if the disagreement relates to the project progress.
8. If the disagreement is personal, set aside the situation from the project environment.
9. If the disagreement is technical, reach an agreement adding value.
10. Keep the focus on optimizing the project objectives (scope, schedule, cost, and quality).
11. Write down the lesson learned to avoid similar conflicts in the future.
12. Communicate the outcomes to the stakeholders.

Problem Solving

Problem-solving is a human process characteristic of finding the appropriate solution to resolve problems. It is a vital attitude when working on projects because people should have the disposition to face issues that are going to show up during the execution of projects. Project people with a positive attitude to solve problems contribute towards delivering successful projects.

Perform an Effective Problem Solving

A technique to apply practical problem solving is to explore options in the gathering information stage. It is imperative to develop options during the analysis of a problem. To do this, first, you must understand the problem and its implications. Then you must apply a technique such as brainstorming, why-why-why, or root cause analysis to identify the possible solutions. As an example, a project conceptual engineering or an option solution study to buy, rent, or refurbish a processing unit should include a decision matrix to display different options with their attributes. To facilitate the decision-making process is essential to include significant attributes associated with the options such as unit capacity, benefits, lifetime, installation costs, maintenance costs, cost savings, among others.

Another effective way to apply the problem-solving attitude is to see the problem from different perspectives, from different points of view. To illustrate this, let us suppose that a project needs to transport high-load process modules from the fabrication shop to the worksite. They have to consider the engineering design (dimensions, weight, packaging), the logistic (moving company, moving media, time of the year), the regulatory requirements (available routes, weight-size limits, notifications), and the public (disruptions during the move). These previous factors are addressed, looking not only at engineering disciplines but also at procurement and regulatory.

Risk is one thing that you need to be proactive with during the problem-solving process. At least a qualitative risk analysis should be performed for an effective problem-solving outcome. Each solution should be accompanied by the associated risk and the plan to manage it. Most of the projects fail because of the exclusion of events with a low probability of occurrence and high impact. For example, developing brownfield projects (projects in existing process plants or facilities) requires identifying the low probability and high impact risk associated. Usually, it is more challenging to handle projects on existing installations in comparison to greenfield projects (projects in undisturbed areas).

Negotiation

Negotiation is the process of exchanging information to reach an agreement. Negotiations take place in almost all aspects of our lives. We negotiate with family members, friends, colleagues, coworkers, and strangers. The outcomes of negotiations are going to depend on the ability and preparation of both parties involved.

Effective Negotiation in Project Management

As project managers, we negotiate with our project team to gain their commitment to execute their work and produce the results. We also negotiate with project sponsors and top management to gain their support. Some circumstances demand our intervention to negotiate with other stakeholders, such as government agencies and the public.

Following are several basic principles to be considered during a negotiation process:

1. Research your topic to be prepared.
2. Ask specific questions to find out the interests of the other party and express your interests.
3. Use the applicable standards as a reference to start the discussion and to reach an agreement.
4. Exchange concessions; if you give a concession, get one for you in return.
5. Remember that both parties should gain something in any negotiation.

6. Keep the focus on the processes instead of the persons.
7. Think about the long-term solutions and not only in the short-term ones.
8. Write down to register the outcomes of the negotiation.
9. Communicate the outcomes to the stakeholders.
10. Compliance with the agreement.

In future project negotiations, apply the previous principles to achieve outcomes towards the project-specific objectives (scope, schedule, cost, and quality), and things are going to start falling into place. Negotiations are not about taking over; they are about reaching consensus to move the project forward.

Leadership

Leadership is the process of influencing others to achieve common goals. In projects, the project manager has formal leadership responsibility, and the project team members have informal leadership responsibilities. The latter is important because everyone should be a leader in their field of knowledge to influence others to cooperate to achieve the project results under his or her responsibility.

Five Principles to Increase Your Leadership Awareness

First principle: Be prepared, never stop learning. You need to grow your confidence and be consistent in your actions, to lead by example. Also, it is essential to create a personal balance lifestyle to be more productive and efficient. Besides working on the project, find some complementary activities such as doing exercises, performing outdoor activities, reading, enjoying art, listening to music, among others.

Second principle: Focus on getting the project results. Discuss possible solutions when facing challenging situations. As an example, I tell my project team that "my door is always open, but do not come to me with just a question, also bring all the options that you can think of." When people show up, we discuss all possible solutions and determine the right one to follow. Leadership is about transforming threats into opportunities and weaknesses into strengths.

Third principle: Pursue simplicity. To illustrate this, when you have a project, your project team list all the activities and include them in a colorful schedule, hundreds of activities. Then when the project starts, everyone starts executing activities quickly, and one month later, you realize that the project is behind schedule. Instead, as leaders, focus and work on your project deliverables and work packages, and act, don't be afraid of taking calculated risks.

Fourth principle: Communicate the meaning of your work. When interacting with others, to gain their cooperation, it is imperative to communicate why you are working on a particular scope of work and describe the interconnection of your work with the overall project work plan.

Fifth principle: Support your project team. For example, project managers should give on-time recognition for coworker's contributions, and people should gain coworkers' trust being supportive. Pay attention to the details when dealing with power skills, because the difference between average and excellence here resides in the details.

Alignment

Project alignment has two dimensions in the project community. The first one is the alignment of the project to the business objectives, strategies, mission, and vision of the organization. The second one is the condition that each stakeholder should be on the same page regarding the main project objectives (scope, schedule, cost, and quality). Both definitions are essential to ensure not only the success of projects but also to obtain the project benefit realization.

Project Alignment with The Vision & Mission and Alignment Within Itself

The vision is like a picture of the organization in the long term (e.g., 5 or 6 years) it presents the status (how and where) we would like to see our business in the future. The mission is a mandate statement that does not change over time. The driving forces that guide us toward our vision and mission are the business strategies and objectives. Business strategies are long term plans (2 or 3 years), and business objectives are measurable short-term plans (usually 1 year). Therefore, projects are in alignment with the vision when they contribute to achieving the business objectives and strategies. As an example, let us suppose we are in the power generation business.

- Our vision reads: "To grow and become the largest high-quality power generation company in the region."
- Our mission reads: "To generate power with the lowest impact on the environment."
- As a result, one of our strategies could read: "To expand wind and solar power generation capacity."
- And one of our objectives could read: "To increase 10% wind and solar power generation capacity this year."

The alignment among the stakeholders with the project objectives is a challenging condition to achieve because of current not integrated project management practices. To illustrate this, during the planning phase, the following steps; scope definition, schedule elaboration, and cost estimates are not consistent because they are usually performed in sequence and by different entities. Besides, during the execution phase, it is also difficult to keep consistency among the scope, schedule, and cost when changes happen, making it challenging to determine the real status of the project at some point in time. As a result, not everyone is on the same page, to avoid this lack of alignment, it is crucial to implement more integrated project management approaches to improve communications. Communication is the most powerful tool to achieve project alignment within all the stakeholders during the project life cycle.

Optimistic Bias

The optimistic bias in projects is the positive perception of the time required to execute an activity or to produce a determined deliverable or work package the first time. And usually, the idea that we can finish something when, for the first time, we say we are going to do it, is an example of the optimistic bias. After you perform something, the next time you are going to do it, you have a point of reference, you have a baseline because you have lived the experience.

The Effect of The Optimistic Bias in Projects

Usually, the optimistic bias harms projects. The positive thinking attitude, most of the time, made us elaborate on unrealistic schedule estimates. To make matters worse, people approve this unrealistic time estimates because they want to be more productive. The problem is that productivity is not a wish, it should be based on facts instead of opinions, and it should be measured. For example, it is commonly accepted to prepare a schedule based on the opinion on how long the activities are going to last, when the time should be the result of dividing the amount of work by the number of available resources, and the latter is the capacity of doing the work involved.

Another example, once we were delivering a training course, and as part of the session, we had to review the 50-question exam from 25 participants. We said to the participants we are going to review the exams in one hour, and after that, we would continue with the training session. It took us 2 hours to complete the entire revision, an example of why an estimated time without any fundamental calculation that involves the capacity of doing the work, is not reliable, so you should avoid this practice as much as possible when preparing your project schedules.

Result-based Concept

Project Management Levels

In general, people do not apply the same project management level regardless of their work experience. Some people handle projects by managing schedules and tracking actual cost versus budget only, and some people know the project management techniques and apply most of them on their projects. The former has a basic project management level, and the latter have and advanced project management level.

Difference Between Basic and Advanced Project Management Levels

A basic level of project management is when during the planning phase, people mainly use a schedule and some documentation to define the work scope. Also, during the planning phase, the cost estimate is determined (and the budget is approved) based on these two previous documents. Usually, these three planning steps (scope, schedule, and cost) are developed in sequence and by different entities. During the execution phase, the schedule is the primary tool to control the project's progress. At this level, the conventional calculation tools are Excel spreadsheets. The problem with this practice is the information update. When changes happen (and they will), it isn't straightforward to incorporate them in the scope, schedule, and cost estimate, and this negatively affects the consistency among these documents and the reports for the management decision-making process, here usually not everyone is on the same page.

An advanced project management level implies the use of several project management techniques and sophisticated software applications. Here the planning process is more elaborated, and people may use earned value management to control the project during the construction phase.

I said "may use earned value," because it is well known in the project management community that this technique is the most powerful one to control the execution of projects, but it is also rarely used worldwide. With the use of sophisticated software, the project team has a better chance to plan the project scope, schedule, and cost in a more integrated fashion. Therefore, during the construction phase, the performance of the project can also be tracked more effectively. The effectiveness of this advanced practice level depends on the appropriate definition of the scope of work and the ability to incorporate changes during the construction phase to maintain the integrity of the baseline plan; here, everyone should be on the same page, from start to finish.

Result-based Concept

The current project practices have a common problem regardless of the project management level. The problem is that people usually focus on activities or tasks, instead of deliverables and work packages. People focus on "doing work" by executing the activities listed in the project schedule, instead of "producing results" that should have been firstly established in a work breakdown structure. A result-based technique changes this and allows focusing on the project results - deliverables and work packages instead of activities and tasks.

The result-based technique shall be implemented during the planning phase while defining "what we are going to do" (the scope of work) and specifically when creating the work breakdown structure. The result-based concept consists of creating the work breakdown structure, breaking down extensive work components into smaller work components until listing the discrete deliverables at each penultimate level, and the work packages (needed to produce the deliverables) at the ultimate level in each work breakdown structure branch. Activities are "how we are going to do it." They are listed for each work package, and they are needed to prepare the project schedule, as such activities are defined in detail to determine the actions, we need to take to create the deliverables and work packages. Later on, during the construction phase, an activity schedule is appropriate for the performers, such as a Sub-trade Contractor, that is building drywall. However, a work plan with deliverables and work packages is appropriate for General Contractors to plan and control the project as a whole. General Contractors need to monitor the progress of construction projects to track margins.

The result-based technique allows focusing and managing the project at a higher level because it is more effective to control 200 deliverables/work packages than to control 2000 activities/tasks. Working on a higher level makes a more straightforward overview of the entire project for effective planning and construction of projects. Practitioners of both project management levels can quickly adopt this result-based concept approach.

Complex Projects

Typically, a complex project involves to study, design and construct process facilities and infrastructure for extreme operating conditions such as high temperatures and pressures (HT/HP), extreme low (-50 C) temperatures, and significant flow rates of solids, liquids, gases, or a combination of them. Usually, the capital investment in complex projects is high.

Examples of Complex Projects

Oil and gas drilling, downhole completions, surface facilities, surface infrastructures, and equipment maintenance are complex projects. They are risky and intensives in capital investment. Some of them involve drilling, mining, extracting, transporting, and processing a large number of hydrocarbons (usually oil, gas, water, and sand mixtures) from wellheads and mines to process and shipment facilities. What is more, the hydrocarbon reservoirs are usually located in remote locations like deserts, jungles, offshore deep-water, and onshore desolated areas.

The capital expenditure in surface process facilities projects is high, specifically for new installations. Let us consider the project sizes according to the total installed costs (TIC), as follows:

- Megaprojects have TIC greater than $1B
- Large projects have TIC between $100M and $1B
- Medium projects have TIC between $1M and $100M
- Small projects have TIC less than $1M

Taking into consideration the previous assumptions; the following are examples of oil and gas surface facilities projects with an idea of their associated order of magnitude capital investment:

1. Design and construction of floating oil production storage and offloading vessels are megaprojects with capital investments usually greater than 10 billion dollars.
2. Greater than 18 inches diameter and more than 1,000 kilometers of oil or gas pipelines usually cost more than 8 billion dollars.
3. A 30,000-barrel oil sands plant project costs around 2 billion dollars.
4. A 10,000-barrel oil and gas flow station facility cost around 150 million dollars, depending on the plant capacity.
5. Oil and gas plant capacity expansions are medium-cost range projects with a final cost between 1 and 100 million dollars, also depending on the work scope modifications.
6. Installing an 8 million standard cubic feet compressor in a gas wellhead is a small project that costs around 0.9 million dollars.
7. Installing an artificial lift method in an oil well is usually a small project ranging between 0.2 and 0.5 million dollars.

INTEGRATED PROJECT PLANNING AND CONSTRUCTION BASED ON RESULTS

There is no doubt that oil and gas surface projects are capital intensive, and the size factor increases the risk of failure because statistics show that the larger the project, the higher the percentages of cost overrun and schedule slippage during the project construction phase.

The Result-based Concept and Complex Projects

Oil and gas projects are examples of complex projects; they place challenges to be managed properly. As a result, their rate of failure is higher in comparison to other industries. This situation represents an opportunity to implement the result-based concept. Stakeholders can focus on generating the agreed results; project progress is going to be determined if deliverables and work packages are completed. In the end, this is going to increase the accuracy, consistency, and predictability of complex projects and, in general, for large, medium, and small projects as well.

Project Planning Phase and Result-based Concept

The Scope Definition

The scope definition is the planning step process executed by the project team to determine what is going to be produced during the construction phase. Usually, a project starts with a high-level scope of what is expected. Depending on the stage of development, this high-level scope is written in the project charter, contract agreement, or engineering document signed-off by the client and provider. During the planning phase, it is imperative that the project team or project performers work together to elaborate a detailed scope of work and determine what they are going to work on in the construction field and finally deliver. It is also useful to identify what is not included in the work scope. This exercise helps the project performers understand the boundaries to avoid scope creep and to ensure that the project is going to deliver a product or service in compliance with the agreed technical and regulatory requirements.

Scope of Work – Work Breakdown Structure – Scope Baseline

The scope of work, work breakdown structure, and scope baseline are terminologies used in the literature to name the same project topic, "what the project team is going to build and deliver." The scope of work is usually a narrative description (known in the literature as the project scope statement) of the work to be performed. The work breakdown structure is a map that displays in a hierarchical chart the work scope to be delivered. And the scope baseline consists of these two documents.

The easiest and practical way to visualize and elaborate the work scope is to represent it in a work breakdown structure, which is a hierarchy diagram or a tree chart. Nouns and adjectives are used to name their work components, deliverables, and work packages, and it shall be developed before creating the project schedule. As an example, the following hierarchy diagram shown in figure 2, is a result-based work breakdown structure of a House Construction project; it was created, listing the deliverables at the penultimate level and the work packages (to produce the deliverables) at the ultimate level of each work breakdown structure branch.

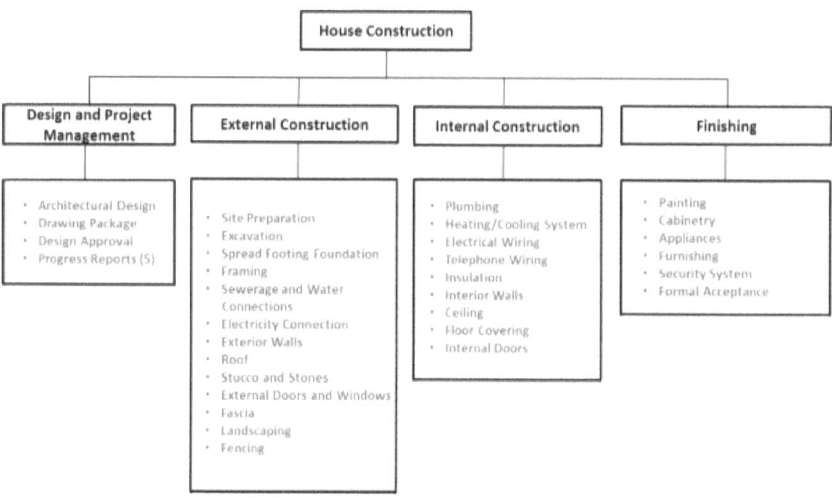

Figure 2. Work Breakdown Structure Based on Deliverables and Work Packages

The work breakdown structure is created to measure the project performance regarding the scope of work. Since evaluation is a comparison, the measurement of project performance in the scope dimension, is the comparison between the scope baseline represented by the work breakdown structure, and the percent complete of each deliverable, as a function of the work packages finished according to the quality plan, during the construction phase.

INTEGRATED PROJECT PLANNING AND CONSTRUCTION
BASED ON RESULTS

The work breakdown structure contains the entire work scope. A scope baseline based on results - deliverables and work packages - keeps the stakeholders focused on the most important objective; to produce the agreed project results according to the quality plan, this should be accomplished in an easy to follow, yet powerful way.

Resource Allocation

Resource allocation is the process of assigning items such as labor, materials, equipment, office space, contractors, sub-contractors, procedures, specifications, and others to the project, to produce the agreed results (deliverables and work packages). The resource allocation is the process that allows determining a consistent and realistic project integrated baseline (integration of scope baseline, schedule baseline, and cost baseline) because the allocation process takes place during the planning phase, and it is implemented during the construction phase of projects.

Effective Resource Allocation

For consistency, the resource allocation starts during the initiating and planning phases. Here the project team or project performers work on cyclical iterations between the work plan and the scope baseline to determine the number of resources required to produce the deliverables and work packages within the timeline of the project and in compliance with the quality plan. In the planning phase, the project performers specifically need to check the availability of the resources. They also need to start earlier negotiations to compromise critical resources, for example, to engage subject matter experts, to book heavy equipment during plant shutdown periods, or to buy software licenses. During the construction phase, the resource allocation takes place, usually with mobilization to the field of construction, office allocations to the project team, issuance of contracts to service and material providers, buying or renting office equipment and machinery, buying material, buying software licenses, distribution of procedures and quality specifications, distribution of the baseline plan (for scope, schedule, and cost), among others.

Result-based Increases the Consistency of Resource Allocation

One of the critical factors to ensure the execution of the project baseline plan is resource availability (as established during the planning phase) to perform the work according to the project objectives. The result-based concept facilitates resource allocation because resources are estimated, focusing on producing the agreed deliverables (D) and work packages (WP), instead of the currently common practice of allocating resources to activities or tasks. Sometimes, for no-loaded schedules, these allocations never happen or are deficient. The result-based concept requires a change in the way of thinking about projects, instead of thinking about activities, it is imperative to shift the thinking towards results. As an example, in the House Construction Project, the resources allocation of each deliverable is as follows:

- D: Design and Project Management. WP: architectural design, drawing package, design approval, and 5 progress reports. Resources: construction manager, architect, civil/structural engineer, drafter, foundations inspector, rough framing inspector, plumbing inspector, electrical inspector, and final approval inspector.
- D: External Construction. WP: site preparation, excavation, spread footing foundation, sewerage and water connections, electricity connection, framing, exterior walls, external doors and windows, roof, stucco and stones, fascia, landscaping, and fencing. Resources: superintendent, operator 1, operator 2, helper, backhoe, bulldozer, dump truck, and external construction contractor.

- D: Internal Construction. WP: plumbing, heating/cooling system, electrical wiring, telephone wiring, insulation, interior walls, ceiling, floor covering, and internal doors. Resources: internal construction contractor.
- D: Finishing. WP: painting, cabinetry, appliances, furnishing, security system, and formal acceptance. Resources: finishing contractor.

This approach of determining and allocating the resources focusing on deliverables and work packages (instead of activities or tasks) is a more appropriate and practical approach to make a project successful and contribute to building accuracy, transparency, and consistency during construction.

Cost and Schedule Estimates

Cost and Schedule estimates are iterative and cyclical project planning stages where resources such as labor, materials, equipment, contractors, and sub-contractors are determined consistently with the project results previously established in the scope baseline or work breakdown structure. This calculation process needs to be in alignment not only with the project objectives (expected timeline, quality requirements of products or services, and expected investment) but also with the resource availability to obtain a realistic calculation output. The outputs of the resource estimate process are a cost estimate, a schedule estimate, or a work plan.

How to Determine Realistic Cost and Schedule Estimates?

Once the work scope is established, the estimation of the cost and schedule needs to be determined with the participation of the project team and other influential stakeholders (e.g., project sponsor, contractors, sub-contractors, service providers, subject matter experts, or final user). In the House Construction project example, each specific work breakdown structure deliverable and work package needs to be further defined with a specific description and due date. The latter depends on the required and available resources such as construction manager, architect, civil/structural engineer, drafter, inspectors, superintendent, operators, helpers, backhoe, bulldozer, dump truck, contractors, or sub-contractors. Only the participation and commitment of everybody involved makes cost estimate and schedule estimate feasible and realistic. Finally, a contingency or percentage of the final estimated cost and schedule is recommended to be included in the estimates to account for the identified risks that may happen during construction due to the complex nature of projects. The cost and schedule contingency percentage numbers can be determined after performing a project risk assessment. It can also be defined by the lessons learned from previous projects (e.g., usually a number lower than 10% for construction projects), or can be a combination of both a risk assessment and the lessons learned. In the end, it is better to have some contingency reserve than have nothing. Because going to the field of construction with no contingency means that if something happens due to uncertainties (e.g., procurement delays, rework, or bad weather), the project can quickly become over budget and behind schedule.

Result-based and the Accuracy of Cost and Schedule Estimates

A scope definition based on results (deliverables and work packages), allows determining more objectively the project baseline plan as a function of the expected completion date of work components during the project timeline. For the House Construction project example, the project plan is established assigning work breakdown structure deliverables and their work packages to control accounts or cost centers, and estimating the cost of the resources required to produce these planned deliverables and work packages during construction. After estimating the resources, the due dates of work packages and deliverables are determined, because the schedule depends on the availability of resources to perform the work to produce the project results. The project work plan is developed by an iterative process; at this stage, resources are allocated in agreement with the convenient due dates, proportional to the amount of work involved, and depending on their availability.

This approach also supports the establishment of a consistent, integrated time-phased budget baseline, also known as the s-curve or integrated performance baseline, once the cost and schedule estimates are approved. Using a project work plan based on results (deliverables and work packages) is crucial because, currently, companies want to know the real project status based on facts instead of opinions about the percent complete of activities or tasks. Using a work plan based on results, as time goes by, during the project execution, the status of deliverables and work packages are verified with a "Yes/No" question posted to the project performers and easily verifiable in the field of construction with inspections according to the quality plan. The questions and inspections reduce the subjectivity when reporting project progress, facilitates the correct use of earned value, and improves the overall project performance measurement process. Increasing the quality of the estimates is a golden rule to establish a more realistic and feasible baseline plan to increase accuracy, transparency, and predictability in projects.

Importance of Establishing A Baseline Plan

A baseline plan is the product of the integration of scope, schedule, and budget (or approved cost estimate). It is a fundamental step that defines how the project is going to be executed from the start (point A) to finish (point B) in the field of construction. Under the result-based concept, the scope is established in a consistent work breakdown structure, the schedule is the sequence of deliverables, and work packages (agreed results) and the budget is the time-phased approved estimated cost that contains quantities of resources within time. This plan is established to measure project performance. Since evaluation is a comparison, the measurement of project performance is the comparison between the baseline plan and the actual work accomplished; this is the reason why the baseline plan is also called the performance measurement baseline.

Determine A Realistic Baseline Plan

Currently, the planning process, scope definition, schedule definition, and cost estimate are usually developed in sequence by different entities, what is more, the cost estimate is not distributed within time. And changes are not captured and incorporated into the plan consistently when they happen during construction. These factors are the main reason why usually the planning process is not always consistent, and this is one of the reasons why projects fail. It is necessary, first, to define the entire scope of work in a consistent work breakdown structure listing all the results at the penultimate (deliverables) and ultimate (work packages) level of each branch. Second, to estimate the resources and time required to produce the results with the involvement of people that are going to build the project. And third, to discuss and approve the work plan by the main stakeholders (e.g., project sponsor, top management, or client) to develop a realistic baseline plan to increase accuracy, transparency, and predictability.

How Does Result-based Increase the Consistency of The Baseline Plan?

The practical planning approach consists of creating the work breakdown structure using the result-based concept (scope baseline) and developing a work plan listing the control accounts or cost centers with their deliverables and work packages within time (schedule baseline). And determining the cost of resources required (cost baseline) to produce a consistent, integrated time-phased budget baseline once the work plan is approved. As an example, in the House Construction project, four deliverables and 36 work packages were defined (Figure 2). If we place each deliverable with its work packages in a control account or cost center (to be able to control the percent complete of each deliverable), the result is four control accounts in a timeline of e.g., five months of construction. Then, the cost is calculated for categories such as labor, material, equipment, contractor, and sub-contractor during the time frame execution of five months. The cumulative sum of the estimated daily cost of all control accounts or cost centers within time generates the cost estimate curve at the project level. The approved cost estimate curve or project budget, as a function of time, represents the planned value curve, performance measurement baseline, s-curve, budget baseline, project baseline, or the baseline.

It is imperative to have a baseline before going into the field of construction. The baseline is the best estimate of how the project is going to behave during construction. In construction, project management is like conducting a vessel to its destination on time, on budget, and specs regardless of the obstacles, and to perform this, the project team needs a realistic navigating map, which is the project baseline.

Project Construction Phase and Result-based Concept

Procurement

Procurement is the process of acquiring resources for projects; labor, materials, equipment, contractors, and subcontractors, thorough purchase orders or service contracts. During the stage development sequence of projects, procurement comes after engineering and before fabrication or construction. Both purchase orders and service contracts are written binding legal agreements between providers and clients.

Effective Procurement

Effective procurement is critical to ensure the success of projects. Procurement should start during the planning stage with the definition of the procurement strategy. The procurement strategy is part of the project execution plan. It establishes the resources to be purchased, location of potential providers, suppliers list, contracting plan (fixed price, cost reimbursable or time and material), quality assurance through third-party inspections, and the procurement process itself like the bid process and award steps.

During the project execution, once the design specification or basic engineering is developed, it is appropriate to procure the material with long-lead delivery time, such as more than 6 months. Examples of long-lead items are large size valves, pumps, boilers, storage tanks, compressors, and line pipe. During detailed engineering, technical information is taken from vendor drawings and included in the final issue for construction drawing package by disciplines such as civil and structural, mechanical, electrical, and automation and control.

The procurement process can be described at a high level as sending out the material or contract requisition, evaluating the quotes (for materials) and bids (for services) from different potential suppliers, and awarding sending the purchase orders for materials to the seller, and signing the contracts for services with the selected provider.

It is essential to select good service companies to work with and provide them the appropriate work scope information. Also, the project should engage suppliers at the very early stage of the project. The idea is to establish a trustworthy business relationship because, in the end, the suppliers are part of your project team. You should see your suppliers and service providers as an extension of your project team; the project success depends on their quality contribution of materials, equipment, and services. As an example, fabrication at the shop should start once all materials are delivered to avoid costs related to shopping machinery set up or rework due to changes. The progress of construction also depends on contractors, their availability, and on-time supply of materials, equipment, and skilled people at the site. Construction contractors also supply consumables, essential items related to the quality of constructions in general (e.g., paints, coating, welding rods, sand for sandblasting, among others).

Result-based Concept and Procurement

The result-based concept contributes to establishing effective communication with vendors and suppliers because of the requirement of services and materials can be easily conveyed based on project deliverables and work packages. If everyone uses the same elements of communication (e.g., a scope of work defined with deliverables and work packages), then we are going to be able to minimize clarifications and changes during the procurement process. Because with an approach based on results, everyone is going to be on the same procurement page.

The Project Construction Phase

The project construction starts once the baseline plan is approved and published. The planning process is a preliminary stage to prepare and approved the baseline plan. After the approval, the baseline plan is frozen, published, and sent out to all stakeholders. This step is crucial to start the construction phase because one of the reasons why projects fail is due to the lack of effective communication. The publication of the baseline plan also communicates the agreed construction sequence and expectations and makes the alignment of stakeholders stronger from the beginning.

Things to Focus During the Construction Phase

During the construction phase events take place as follows:

1. Field resource allocation; assignment of project field team members, allocation of field office space and equipment, purchase of material, sign-off contracts, among others.
2. Field resource interaction to produce the project results - deliverables and work packages.
3. Quality checking of products, services, deliverables, and work packages.
4. Track and record actual costs.
5. Meetings among stakeholders, project team, service providers, clients, government, or the public.

The construction project status report is created and communicated periodically according to the reporting periods (e.g., weekly, or monthly). Currently, the activity/task percent complete is a common practice that incorporates subjectivity in the measurement of project progress during the construction phase. It is more effective to use a result-based (deliverables and work packages) baseline plan to keep all stakeholders focused on the most critical thing, which is to produce the agreed project results.

Result-based Concept Increases the Probability of Project Success

As an example, during the execution of the House Construction project, the construction manager and sponsor is going to obtain and allocate the resources (project team, office space, materials, equipment, contractors, procedures, or specifications) to focus on producing the deliverables and work packages which are:

- Design and Project Management with the work packages; architectural design, drawing package, design approval, and 5 progress reports.
- External construction with the work packages; site preparation, excavation, spread footing foundation, sewerage-water connections, electricity connection, framing, exterior walls, external doors-windows, roof, stucco-stones, fascia, landscaping, and fencing.
- Internal construction with the work packages; plumbing, heating/cooling system, electrical wiring, telephone wiring, insulation, interior walls, ceiling, floor covering, and internal doors.
- Finishing with the work packages; painting, cabinetry, appliances, furnishing, security system, and formal acceptance.

The completion of these deliverables and work packages as planned during the timeframe of 5 months ensures that the project is progressing as planned. In addition, the result-based approach facilitates the implementation of earned value and earned schedule because project progress is credited only if deliverables and work packages are completed and as a result, this approach increases the objectivity and probability of project success during the construction phase.

Earned Value

Earned value is the value of the asset that you get after spending or investing resources during the construction phase of projects. It is the most powerful technique to determine the current status of projects because it is based on the physical work completed. A comparison between the resources planned to be spent (budget) and the resources spent (actual cost) is not sufficient to determine the project status and performance. It is necessary to include in the comparison the actual work accomplished to understand the level of completion of a project.

Earned Value Definition

Let us use the 10-10-10 Project example to help understand the definition of earned value. The project consists of building a 10-kilometer road. The project team is planning to build the road in 10 months and spend 10 million dollars. Let us suppose that the project team is going to build 1 kilometer of road each month, and they are going to spend 1 million dollars each month. Although roads are built differently, use your imagination for a moment to understand the concept of earned value. As a result, the project baseline is a 45 degrees straight line, as shown in figure 3, the project is going to spend 10 million dollars in 10 months, to build 10 kilometers of road.

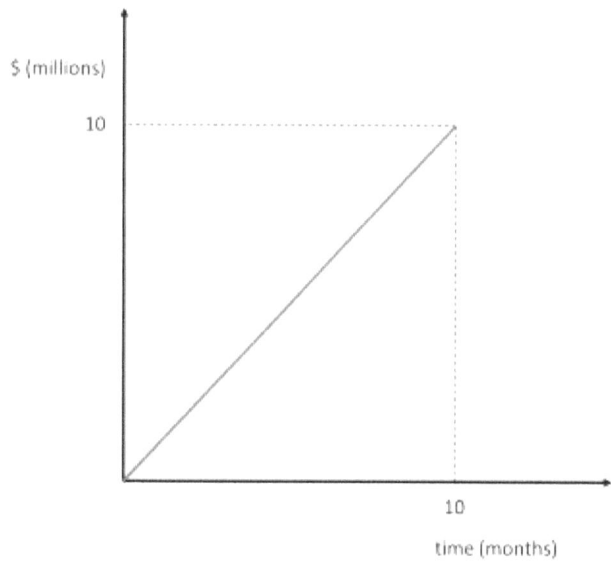

Figure 3. The Baseline of the 10-10-10 Project

Construction starts, and after 3 months, the planned value is 3 million dollars, and the actual cost is 3.05 million dollars. Looking at these two numbers, it seems that the project is going well. However, the amount of road built is 2.5 kilometers, which means that the earned value is 2.5 million dollars, 0.5 million dollars lower than the planned value, and 0.55 million dollars lower than the actual cost. The project is not going well because the planned value ($3M) is money in the bank (ready to be spent by the project), the actual cost ($3.05M) is the amount of money paid to the contractor, and the earned value ($2.5M) is the number of assets built. A project is the transformation of money in the bank into assets, another definition of what a project is.

Use of Earned Value

INTEGRATED PROJECT PLANNING AND CONSTRUCTION BASED ON RESULTS

Even though the earned value is the most effective way to measure project performance, it has not been widely applied or used worldwide. One of the reasons is the current practice to determine the project progress through the activity/task percent complete, which is not only an indirect method to determine real completion, but also incorporates subjectivity in the measurements. As an example, in the House Construction project, under the current practices, the project team list all the activities or tasks required to finish the project. Then the project team includes them in a schedule; once the execution starts, the project team estimates the percentage of the work completed indirectly against the completion of each activity/task in the project schedule. If the project team changes their way of thinking and focuses instead, on deliverables and work packages required to elaborate the House Construction project. Then, the project team would be able to determine more objectively the project progress as a function of the work packages completed at some point in time. Currently, companies want to know the real project progress and status base on facts instead of opinions. Using a project plan based on results is going to facilitate the implementation of earned value because people are going to be able to directly measure the physical work completed more efficiently according to the quality plan.

Implementing Earned Value

The earned value uses several techniques to give credit and measure the project progress based on the work performed in a reporting period. The literature includes the following techniques:

- Fixed Formula: two percentages are assigned for short duration tasks/activities, one when the work begins and the other when the work is completed.
- Weighted milestone: it assigns a percent complete based on a predetermined value for each milestone achieved.

- Percent Complete: the performer assigns an estimated percentage of completion to each activity/task.
- Level of Effort: an automatic value is credited at each reporting period for each task/activity.

These techniques are implemented working with activity/task-based methods, and they add complexity during the implementation of earned value. Because it is more difficult to control, for example, a schedule of 2000 activities than to control a work plan with 200 work packages (and 20 deliverables), all of these tools relate to the same project. A simplified and practical approach to help implements earned value is to credit value only if the work has been completed. For the House Construction project, the project team is going to credit the value that the project obtained (planned value associated) only for the work packages that have been completed to build the house. This approach is a simplified and practical application that can increase the correct use of the earned value technique and the accuracy, consistency, and transparency during construction.

Earned Schedule

The earned schedule is a time-based approach introduced by Walt Lipke in his internet article "Schedule Is Different" in 2003. It is the cumulative scheduled time earned consistent with the work performed or the earned value of a current baseline plan to forecast the completion date of projects. Its application is directly related to earned value and perfectly complements the approach to determine the current status of projects. With earned value, the project team can determine the project status and performance in a work-based dimension. However, it is necessary to incorporate an earned schedule to be able to understand the time-based dimension during the construction phase of projects.

Use of Earned Schedule

Although the use of earned value is growing worldwide, specifically in large and megaprojects because it provides early warnings helping to achieve cost and schedule goals, for fixed-price contracts, the application of earned value is discouraged. Usually, customers believe that since they are transferring the risk of cost to the contractor, then there is no need to implement earned value. Here is when earned schedule acquires importance because the earned value can be applied on fixed-price contracts using the earned schedule concept, even though the cost component is less important for the customer, the time or schedule component becomes a priority. Actually, earned schedule and therefore earned value should be applied in all projects regardless of the contract type (fixed-price, cost reimbursable, or time and material). Because although the contractor may not need to share the cost performance on fixed-price contracts, they should share the schedule performance and forecast using the earned schedule technique. The information about the physical work completed and when the remaining work is going to be performed is still essential for both customers and contractors. Regardless of the type of contract in place, this information is essential during the construction of projects.

Using the earned schedule, the project team can forecast the completion date of projects according to the current progress rate and performance up to date.

Effective Project Reports

A useful project report is the one that periodically conveys to critical stakeholders, the status of the project in a direct and easy to read way. Once the frequency is established (e.g., weekly, monthly, or the last Friday of each month), the report is going to be issued as planned.

The Importance of Accurate Project Status Reports

One of the main reasons why projects fail is because of poor communications, and the project status report is the tool that project managers must use to proactively overcome this issue. Project reports are essential because their content allows key stakeholders to think and make decisions needed to move the project forward.

The type of content of a project status report needs to be agreed with the stakeholders at the beginning. In general, the content answers the most important common questions among the stakeholders:

Where are we?

Here, the report should contain the percentage of project completion, and if the project is ahead or behind schedule and under or over budget. The content needs to display the time and cost planned and incurred up to date.

Where are we heading?

It is related to the forecast of the completion date and final cost. The effective way to forecast cost is by using earned value, and the only way to forecast completion date is by using earned schedule.

How are we doing?

In this part, project managers inform how the project is using resources and time. It is easily presented using ratios or indices, such as the cost performance index (CPI) and the schedule performance index (SPI), which are the best productivity and efficiency indices, respectively.

How are we going to address the main issues?

Briefly, the report should state the action plan to address significant issues in the horizon, because the only way to overcome issues is with the commitment of the involved stakeholders.

Change Management

Baseline Plan Versus Actual Field Data

Once the project Baseline Plan has been published, the construction phase of the project shall start. At this point and according to the result-based concept discussed in previous chapters, the interaction of resources, elaboration of deliverables and work packages, and tracking of actual costs take place. The measurement of the project performance during the construction phase is the result of a comparison among the baseline plan, physical work accomplished, and the cost incurred. It is generally accepted that earned value is the most effective way to measure project performance. Earned value management measures project performance through three usually monetary variables: Planned Value (PV), Earned Value (EV), and Actual Cost (AC). Here, Planned Value is the baseline plan in the form of money in the bank, Earned Value is the work accomplished in the form of assets, and Actual Cost is the cost incurred through the invoices the project has paid or received during construction, at some point in time. Cumulative values of these three monetary variables are used to forecast the final cost and completion date of the project, based on past performance, and each time a reporting period is closed.

Definition of Variances

Earned value management uses the following main project performance metric variances:

a) Cost Variance (CV): determines whether a project is under, on, or over budget. The equation is $CV = EV - AC$. Positive values mean the project is under budget and vice versa.

b) Schedule Variance (SV): determines whether a project is ahead, on or behind schedule. The equation is $SV = EV - PV$. Positive values mean the project is ahead of schedule and vice versa.

c) Variance at Completion (VAC): indicates how much over, on or under budget, the project is going to finish if current performance continues. The equation is $VAC = BAC - EAC$ (where BAC is Budget at Completion and EAC is the Estimate at Completion). Positive values mean the project is going to finish under budget and vice versa.

Managing the Variances

There is a high probability that variances are going to occur during the construction phase because projects are complex, and they involve a dynamic integration of scope and resources within time to produce the deliverables and work packages. There is nothing wrong with the variances themselves; the critical issue here is to identify them and take appropriate corrective actions. Earned value calculations present the variances per control account (or cost center) and per the entire project, for specific reporting periods. Variances can be addressed using the management by exception criterion to determine the root causes focusing on the control accounts or cost centers that show high absolute variance values. According to Pareto's principle, look for the few reasons that cause significant deviations to the expected results. It is also essential to develop recovery action plans to respond to the control accounts or cost centers that show more negative variances. Through management by exception, the project manager tracks and reviews the variances but does not intervene unless corrective actions are necessary, e.g., if the variance is higher or lower than a predetermined percentage value (e.g., +/- 20%).

Project Changes Will Occur

Project management is the process of handling resources to generate specific products or services. When implementing the result-based concept, these specific project outcomes are deliverables and work packages that were agreed during the negotiation and requirement definition stage. Projects involve a dynamic integration of scope, time, and budget according to quality, risk, and stakeholder expectations; therefore, it is almost sure that changes are going to happen during the construction phase. As project leaders, it is vital to accept this fact, be prepared, and focus on how to address and handle the changes.

Project Constraints and Change Management

The project constraints model is used to understand the impact when changes arise during construction. The graphical representation of this model is with a hexagon formed with the project elements: scope, resources, schedule, risks, budget, and quality one on each side of the hexagon. The interpretation of how it works is as follows; if one element changes, then at least one of the other elements is going to change accordingly to maintain the hexagon. Let us review some examples:

- If you add more scope of work to execute, then this is going to be reflected in more budget or more construction time.
- If you accelerate the speed of construction to reduce the schedule, then you are going to have to increase the budget or affect quality.
- If you reduce the budget, then you are going to impact the amount of work to be completed or limit the execution time or quality as well.

When Is a Re-baseline Needed?

If your project needs to produce a few more deliverables or work packages, you probably do not need to create a re-baseline plan. Few more or fewer deliverables or work packages can be documented, approved, and captured in the actual cost, to explain the variance that you are going to see. A re-baseline plan is required if the current baseline does not represent the way the project is being built, for instance; if the current baseline reflects poor planning and cannot be used to measure future performance, or if the past performance has been different than expected and does not correlate with the existing baseline. In the House Construction project as an example, there are 4 deliverables and 36 work packages, if, during the Internal Construction deliverable you need to install an additional work package such as electrical wiring, you are probably ok. However, if you need to install a new electrical transformer to provide electricity to the house, which may be considered a deliverable, then you probably have to re-baseline the project because the construction is considerably different from the baseline plan. The following general steps are recommended to address and handle project changes:

1. Identify the change proactively.
2. Determine the impact on other elements (scope, schedule, cost, quality, or other).
3. Approve the change.
4. Create a new baseline for future construction work.
5. Communicate the change.

A re-baseline is a change in the current baseline plan for future construction work. The purpose of this change is to correct the main project objectives; scope, schedule, cost, or quality. A re-baseline is implemented after reviewing the past project performance and realizing that, for a reason, what is happening in the field of construction, it is not what we planned to build, or it requires adjustments or modifications.

It is almost certain that changes are going to happen during construction. The project team needs to realize they have to handle the changes, and a re-baseline development is part of the recovery plan to include approved changes to continue executing the project's future work.

When changes are significant as per the new electrical transformer in the House Construction project example, there are three primary reasons to re-baseline the project construction phase, as follows:

1. When a project constraint element changes significantly and at least one of the others changes as well.
2. When the current baseline reflects poor planning and cannot be used to measure future performance.
3. When the past performance has been poor and does not correlate with the current baseline.

An effective practice to establish a re-baseline is to modify only the work remaining, create a new baseline for the future work that starts on the cumulative earned value, and keep the cumulative actual cost independent because it is the same construction project.

Risk Management

Risk is the product of the probability of occurrence of one event times the total monetary impact (negative or positive) of that event. As a result, risk management is the practice of identifying, analyzing, and handling the consequences if an event occurs over time in the field of construction. Risk management is a good practice that must be performed during each project development stage (e.g., study, engineering, or construction) to increase the probability of project success.

A Practical Risk Management Approach

The methodology used to identify the risk events depends on the level of detailed definition and phase of completion of a project. Low detailed scope definition and early phase of completion (e.g., study or conceptual design) indicate that we can apply techniques such as; brainstorming, cause-effect diagrams, interviews, fault tree, or what-if analysis. While a medium to high detailed work scope and a medium to late phase of completion (e.g., basic engineering, detailed engineering, or construction) indicate that a HAZOP (Hazard and Operability Study) is required. During construction, the safety, health, and environment plan must be followed to control risks in the construction site. During the risk identification process, a risk register is created to record, control, and follow-up the risk management process during the project timeframe.

After the risk identification, the next step to manage risks is to analyze them. The purpose of the analysis is to assign a probability of occurrence of the trigger event. This probability could be based on qualitative or quantitative techniques. Most of the time, a qualitative technique consisting of a scale of a high, medium, and low are enough to determine the probability of occurrences and the level of the monetary impacts on projects. Quantitative techniques involved the use of probabilistic analysis, e.g., Monte Carlo simulation, decision tree analysis, or expected monetary value calculations. At least a qualitative analysis should be performed in all construction projects. A matrix of the probability of occurrence versus monetary impact is prepared to analyze and evaluate each risk on the risk register.

The different risks can be prioritized in the risk register with a scale of a high, medium, and low risk as an example, then a decision needs to be made in how to handle each risk depending on the ratings. The strategies to handle negative risks are as follows:

1. Avoid: eliminate the cause of the risk (e.g., remove it from the scope of work).
2. Transfer: allocate the risk to a third party (e.g., outsourcing).

3. Mitigate: reduce either the impact or the probability associate (e.g., add a layer of protection in a safety system).
4. Escalate: escalate risks that are better managed at the corporate level (e.g., insurance).
5. Accept: do nothing but set aside some resources just in case.

When the risk effect is positive, an opposite strategy is appropriate in each case because positive impacts are opportunities. When using opposites strategies, we want to increase the probability of occurrence of opportunities. The opposite strategies are as follows; exploit (to avoid), enhance (to mitigate) and share (to transfer), escalate, and accept are the same for risks and opportunities.

Ending the Construction Project

The Project Completion Stage

During the construction phase, the level of completion of projects is directly related to the product or service to be delivered and the compliance with the agreed requirements. What the project is going to deliver is defined in the scope baseline or work breakdown structure. The agreed characteristics and properties of each work package (and deliverable) determines the quality level that customers and providers are going to check when the work package (or deliverable) is ready. The completion is defined between customers and providers with the agreed characteristics and properties of these discrete project results according to the quality plan.

Requirements of Product and Services

Projects get into trouble mainly because of discrepancies between customers and providers. These discrepancies are related to the compliance of the products and services with the requirements. That is the reason why it is important to establish the common ground of standards for products and services. For example, engineering, procurement, and construction project specifications are usually provided by the customer. These specifications include the industry design standards, codes, and regulations that the company expects its contractor to be in compliance with, for functionality and safety. Also, customers and providers must agree regarding the acceptance criteria of each deliverable and work package.

Increasing the Objectivity on Projects

The result-based concept facilitates and provides maximum objectivity at the moment of reporting physical project status because progress is credited only if deliverables or work packages are completed (instead of measuring the completions of activities or tasks). Therefore, once all the agreed results are finished, and the final project product or service passes the commissioning or quality checking stage, it is considered that the project is complete.

One of the core objectives of project management is to get to the project completion stage, in full compliance with the agreed requirements. Implementing a result-based technique focusing on the project deliverables and works packages and applying a simplified earned value methodology facilitates the achievement of this core business objective.

Commissioning

Commissioning is the process of determining the completion of a project after construction. It is usually performed on each one of the sub-project components such as civil, mechanical, electrical, and instrumentation & control when their construction scopes are finished. It is directly related to quality and stakeholders' requirements. The purpose of this business function is to find out if the product of the project or the ultimate deliverable is correct and complete.

When the project Is finishing

Projects are completed after the commissioning makes the announcement. Usually, the announcement is determined after a series of inspections, checklists, and final verification, as follows:

Civil completion: civil and structural work is generally needed at the beginning of the construction work; therefore, it is one of the first quality control process performed to continue with the construction. It involves earthwork, clearing, excavation, concrete foundations, piling foundations, structural steel, back-filling, building installation, and later landscaping and fencing.

Mechanical completion: it is the verification of the physical installation of equipment and line pipe. For example, compressors, pumps, tanks, columns, towers, separators, scrubbers, centrifuges, valves, dampers, filters, flare stacks, piping, pipelines, risers, air duct-work, insulation, paint, and other protecting coating systems.

Electrical completion: consists of the installation of power supply, cable trays, wiring, heat tracing, lightning, and telecommunication.

Instrumentation & Control completion: here, the following sub-systems are reviewed; control valves, control panels, instrument wiring, instrument air tubing, field instrumentation, and safety & security systems.

Commissioning activities need to be included in the cost estimate and budget because labor, material, equipment, and specialized contractors and subcontractor are going to do the inspections, checklists, and final verification work when appropriate to continue with the start-up of the facility.

Start-up is the process to hand the project over the Operation function for the steady-state functionality and production.

The result-based approach facilitates the commission process because if you plan and execute the project based on deliverables and work packages (instead of activities or tasks), it is easier to control the outcomes when you compare plan versus actual, according to the quality plan. The result-based concept helps to focus on the formal acceptance of the outcome of the project during the commission stage.

Lessons Learned

Lessons learned are the record of positive and negative experiences that happen during the construction of projects. These records are the source of valuable information for future projects to repeat actions to improve project performance or to avoid situations that increase costs or delay the speed of construction. Besides, lessons learned are facts to assess project performance during an audit process. Although the benefits of the lessons learned are well known in the project community, new projects should increase the use of the recorded lessons learned of similar projects, because this is something that still seldom happens in practice.

Recording and Retrieving Lessons Learned

It is important to write down the lessons learned when they occur since everyone is busy working on the project during the planning and construction phases; this is something difficult to achieve. As a result, usually, the recording of lessons learned happens at the end of the project as part of the close-out process. The project manager is the driving force to promote at least one meeting with key team members (e.g., engineering disciplines, buyers, inspectors, construction managers, and other stakeholders) to consolidate and record the critical facts that happened in the project.

An appropriate record of the lessons learned addresses the scope, schedule, cost, and other essential project dimensions such as quality, risk, or procurement. The content of the lessons learned should answer the question, either what did work well? In a positive sense, or what can be improved? If the experience was a negative one. The answers should be structured, taking into consideration the cause and effect analysis, asking "why" at least three times. This technique is used to find out more about something. The record of the date of occurrence is also important for future chronological references.

Lessons learned are placed into the project documentation, in a physical or electronic media. They must be easy to find and access by future projects. Lessons learned are particularly useful during the planning phase of future projects, especially when the project team is creating the work breakdown structure and during the elaboration of the project plans (e.g., quality plan, procurement plan, construction plan, and others).

When implementing the result-based approach, the project team is working at a higher level than the activity or task level, using deliverables and work packages instead. Therefore, it is easier to develop the lessons learned with the cause and effect relationship because everyone is talking at the same level of project information. Projects are complex, and the result-based technique facilitates the implementation of the proven best practice of lessons learned.

Project Planning and Construction Based on Results

Although there is an effort to implement the best project management techniques worldwide, construction projects still experience difficulties meeting the three successful criteria; delivered on time, with a final actual cost on or below budget, and in full compliance with the technical and regulatory requirements.

One of the reasons for the low rate of project success is because the planning and construction phases are not handled in an integrated way, which adds complexity to the challenging project environment. Construction projects are complex; they involve a dynamic integration of scope and resources within time, to achieve business objectives and stakeholder expectations. In general, the larger the project, the higher the percentage of cost overrun and schedule slippage, and nowadays, it is imperative to reverse this trend to maintain a competitive business edge. This scenario is the reason why there is the need for a truly integrated project planning and construction methodology to manage the entire project delivery life cycle from sanction to commissioning, to optimize resources, improve performance, and increase profits.

Another reason for the low success rate is because of the current practice to determine project status with the activity or task percent complete technique. This practice is not only an indirect method to determine physical completion, but also incorporates subjectivity in the measurements because here, the performers estimate the percent complete at each reporting period. To change this practice, project practitioners need to focus on deliverables and work packages (instead of activities and tasks), implementing the result-based concept presented in the previous chapters. The focus on project results increases the objectivity when reporting physical project progress. It improves the accuracy to forecast the final cost and completion date of projects during each measurement reporting period. Also, power skills are more effective in environments with accurate information, transparency, and objectivity that are promoted after focusing on project results. However, the implementation of the result-based concept requires a different way of thinking about projects.

An integrated project planning and construction methodology based on project results combines the scope, schedule, and cost in a baseline plan. This baseline plan allows measuring project performance through the comparison with the physical completion and the actual incurred costs, from the beginning until the end of the construction stage, using consistent project status reports.

In this integrated planning and construction methodology, the modeling of the planning phase starts with the definition of the scope of work with a results-based work breakdown structure, and a work plan and cost estimate based on project results (deliverables and work packages) instead of activities or tasks. During the construction phase, the project performance is measured using earned value and earned schedule, not applying the traditional techniques to credit value. Instead, earned value is credited as work packages (and deliverables) are completed using the binary completion theory. This theory states that something was either done or not done. It is a simplified approach to credit value only for the physical work completed, which increases the objectivity to determine the project percent complete and forecasts of final cost and completion date when closing a measurement reporting period.

The combination of this result-oriented concept with a simplified application of earned value and earned schedule in an integrated project planning and construction methodology increases the probability of success because projects are handled more objectively and proactively, based on performance. This integrated methodology based on results also allows monitoring performance of portfolios and programs more effectively, due to the consistency in recording productivity and efficiency indices (cost performance index and schedule performance index, respectively), project percent complete, and forecasts of final cost and completion date of projects.

This integrated project planning and construction methodology based on results increases the accuracy, transparency, and predictability in construction projects. However, its implementation must be simple for easy adoption into the construction industry. For this reason, a software tool was developed to make possible the implementation of this integrated methodology and the ideas presented in this book in an easy, yet powerful way. The software is a cloud-based application called PEMS (Planning and Execution Methodology Software). PEMS allows the implementation of this integrated project planning and construction methodology based on project results to realize the benefits.

PEMS was developed to encourage the private industry and government organizations to introduce changes in the way construction projects are handled nowadays. The primary drivers to implement an integrated planning and construction methodology based on project results, specifically in the construction industry, are to track margins, avoid claims, and increase profits. Start making a difference in your construction projects implementing this integrated approach based on project results accessing PEMS with a free trial period visiting the following link: **https://www.pems.io/**

About the Author

Williams Chirinos, M.Sc., P.Eng., PMP, co-founder and president of Inexertus, Inc., is a project management consultant with 20+ years of project management accomplishments in planning and execution of complex projects in the Oil and Gas industry and Engineering, Procurement, and Construction Management (EPCM) environments.

He has delivered fundamentals to advanced project management training courses and consulting to establish and operate a project management office (PMO) in several private companies and government organizations.

He is a Project Management Professional (PMP) credential holder and an active PMI member since 2006. He serves as an associate editor for the Society of Petroleum Engineers (SPE) - Production and Operations. Mr. Chirinos can be contacted at williams.chirinos@inexertus.com.

Read more at https://www.pems.io/.